MW01537429

THIS JOURNAL BELONGS TO :

Published By : Journals Factory

Made in the USA
Las Vegas, NV
12 December 2023

82676113R00068